Off Grid Solar Power for Preppers

How to Build, Install and Maintain your Alternative Source of Electricity to Be Safe and Self Sufficient During any Emergency

RAYMOND L. HILLMAN

© Copyright 2022—All rights reserved

This document is geared towards providing exact and reliable information regarding the topic and issue covered. The publication is sold with the idea that the publisher is not required to render accounting, officially permitted, or otherwise qualified services. If advice is necessary, legal or professional, a practiced individual in the profession should be ordered.

In no way is it legal to reproduce, duplicate, or transmit any part of this document in either electronic means or printed format. Re- cording of this publication is strictly prohibited, and any storage of this document is not allowed unless with written permission from the publisher.

Table of Contents

What Is Solar Power? 9
How Does Solar Energy Work? 10
Off-Grid and On-Grid Solar Power: What is the
Difference? 10
Getting Electricity from Off-Grid Solar 10
Grid-Connected Solar Power 11

1. How to Build a DIY Solar Power System 13
Everything You Need to Know Before Installing Your
Solar Power System 13
DIY Installation of a Solar Panel System Step by Step 14
   Mounting the Photovoltaic Installation Brackets 14
   Surface and Placement 14
   Let's Go with the Process of Anchoring to the Solar
   Panel Structures 15
   You Have Another Option 15
   Proceed to Connect the Electrical Inverter 16
   The Inverter 17
   Here Are Some Tips and Recommendations 18
   The Solar Panel Inclination 18
Off-Grid Solar Setup System Components 19

2. How Much Does the Installation Cost? 21
Cost of Solar Panels for Homes 21
Benefits of Solar Panels for Homes 21

Calculating the Cost of Solar Power 22
  Energy Consumption at Home 23
  Calculating an Example So That You Can Apply It 24
  How Many Kwh Does a Solar Panel Produce? 25
  Available Space on Your Home Roof 25
  How Many Solar Panels Do I Need for a 200m2 and 100m2 House? 25
  Example of the Surface Area You Might Require 26

3. Basic Electricity Rules 29
Electricity Basics 29
  Electric Current 30
  Electric Circuit 30
Electrical Magnitudes 30
  Electrical Voltage (U) Or Potential Difference 30
  Electric Current (I) 31
  Joule Effect 31
Resistors 31
  Types of Electrical Resistors 33
Power 33
  A Formula for Calculating Electrical Power 33
Electrical Circuits 34
Types of Connections 34

4. Essential Tools That You Must Always Have Ready 37
Safety Tools 37
  A Safety Helmet 38
  Safety Glasses 39
  Safety Clothing 39
  Safety Belt 40
  Gloves 40
  Safety Shoes 40

Face Protection 41
Power Tools 41
Writing Tools 42
Solar Racking Tools 42
Battery and Maintenance Tools 44

5. How to Calculate Your Electricity Needs 45
How to Calculate Energy Needs 45
Step 1: Calculate the Estimated Consumption 45
Step 2: Calculate the Required Solar Panels 47
Step 3: Battery Capacity 48
Step 4: Selection of Regulators and Converters 50
How to Reduce Electrical Needs 51
Use Less and Better Electricity 51
The Number of Solar Panels You Require 53

6. Building and Installing an Off-Grid Solar System 55
The Best Way to Install Solar Panels 55
How to Wire Your Off-Grid PV Solar System 55
What Are the Special Requirements for Conductors and Cables in Photovoltaic Systems? 56
Why Are LAPP Products Suitable for PV Systems? 57
Building Your Battery House 57

7. Maintenance of an Off-Grid Solar Power System 61
Types of PV Maintenance 61
Factors to Consider When Installing Bunker Panels 62
What do you do with preventive maintenance? 62
Maintenance of Different PV Components 63

How, When, and How Often Do You Start Cleaning
Solar Panels?..63
   Remember not to turn off the system 63
   When and How Often to Clean Solar Panels 64
   Visual Inspection of Photovoltaic Panels......................65
Other Solar Panel Maintenance Tasks 66
   *Tips for Cleaning and Maintaining Solar Power
Over Time* 66
   In terms of frequency 67
   The Step-by-Step Check of the Solar Kit Spaces...............68
Solar Cells...68

8. Solar Panel Elements 71
Solar Charge Controllers 71
   Charge Regulator Characteristics 72
   Types of Charge Controllers 72
   How to Choose a Charge Controller? 74
Solar Battery Bank 75
   Unit of Measurement of Energy Capacity 75
   Depth of Discharge...76
   Life Expectancy..77
Solar Inverters 77
   How the Solar Inverter Works 78
   How are solar inverters classified? 79
   Grid-Connected Inverter 79
   The Inverter for Off-Grid PV Systems................................80
   Hybrid Inverter 80
Conductors and Connectors 80
   Design and Structure..80

9. How to Build Your Solar Panels Safely 85
The Most Important Safety Rules 85
   While Moving It..86

In the Installations 86
During Operation 87
Have Safety Equipment 87
How to Work Safely with Electricity 87
Safety Rules for Working on Roofs 88
Preparing for Bad Weather 89
Lifting and Handling Solar Panels 90

Conclusion 93

What Is Solar Power?

Solar power is one of the most popular and widely used renewable energy sources in our country. At a time when being responsible for our planet and the environment is key to sustainable development, it is important to understand exactly what solar power is, how it works and how we can use it.

Solar energy is the energy produced by the sun that reaches the Earth through radiation. It is a renewable source that human beings seek to obtain efficiently using different technologies that have been developed over time.

This energy source is inexhaustible and very abundant, so in addition to being a renewable energy source, it is also a clean energy source and an alternative to other non-renewable energy sources such as fossil energy or nuclear energy.

How Does Solar Energy Work?

Broadly speaking, once solar energy reaches the Earth's surface, a photovoltaic device is needed to convert it into electricity. The device captures the radiation through photovoltaic cells and converts it into electricity. It is its most common use.

But what happens to the energy before it reaches the Earth? It is created by fusion reactions that take place in the sun. The radiation travels to Earth via electromagnetic waves, which can be used and even stored.

Off-Grid and On-Grid Solar Power: What is the Difference?

Off-grid solar power systems are not connected to the utility grid, while on-grid solar power systems (also known as grid-tied) are coupled to the utility grid. Your choice of off-grid or grid-tied systems will determine your electricity supply, the equipment you need for excess production, what happens when the grid goes down, and how your electricity bills are billed.

Getting Electricity from Off-Grid Solar

What does off-grid solar mean? With an off-grid solar system, you rely entirely on solar energy and energy stored in batteries to power your home or bunker.

If the solar system you choose is not connected to the grid and you do not have a generator, you will only have electricity at two points:

- When the sun is shining, your solar system generates electricity.
- When you get electricity from a solar storage device (such as a battery) that previously produced a solar system.

If you don't have a battery or a way to store energy, you will have less or no power on cloudy days and no power at night.

With an off-grid system, you won't have access to additional electricity if you need it. What you produce and what you store is what drives your equipment.

Grid-Connected Solar Power

This one connects to the grid, but it is not recommended if you plan to have electricity when the world ends. This chapter will tell you all about solar power and how to get it at home by calculating your consumption.

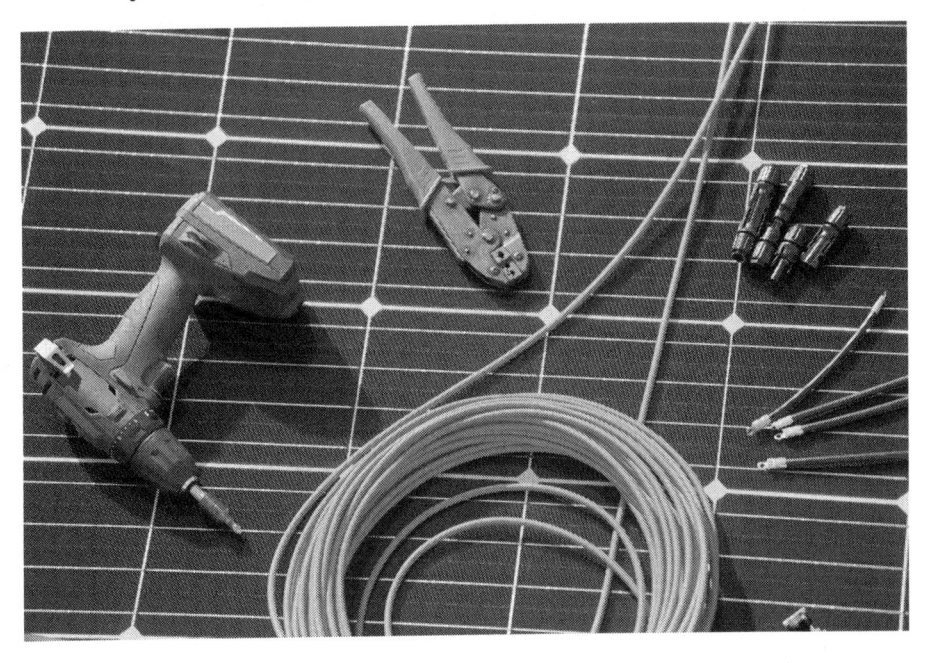

Chapter 1
How to Build a DIY Solar Power System

The first thing you need to know is how to install the system and the steps needed to achieve it.

Everything You Need to Know Before Installing Your Solar Power System

It would help if you took some time to familiarize yourself with the P.V. equipment you want to buy and the quality of its products before committing. Ultimately, the quality of the equipment will greatly affect the performance and longevity of the P.V. system, so it is always a good practice to purchase equipment from reputable and professional manufacturers.

Here are the aspects that you should consider when buying solar panels:

- The brand certifications.
- The quality of the manufacturer.
- Even if you do not enjoy it, their warranty guarantees the longevity of the equipment.
- The visual inspection of the panel.
- Positive power tolerance.
- The efficiency of the solar panel.

The following are steps that should be taken when installing solar panels, although it is advised that you take the precaution of having professional help.

DIY Installation of a Solar Panel System Step by Step

These are the steps you have to follow when you have the system.

Mounting the Photovoltaic Installation Brackets

Before placing the solar panels on the roof, it is necessary to position the structure; depending on the type and slope of the roof, this will be the type of structure to be used. In other words, the structure on the roof with which the coplanar modules are placed will be different from a flat roof where the modules must be inclined to obtain the system's maximum efficiency.

Surface and Placement

Solar P.V. panels can be installed almost anywhere, and often the most suitable options are roofs or floors as they help resist high winds, seismic events, shocks, external forces, and more. Keep in mind that the surface on which you will place the structure must provide:

- an installation that is protected to withstand adversity.
- security and service.
- the optimum orientation and inclination for the application.

Let's Go with the Process of Anchoring to the Solar Panel Structures

After you select the structure, you can start the installation:

- Start by piercing the surface two to three inches to allow you to insert the expander.
- Place the liquid sealant on the perforation to avoid leaks, drips, and more.
- Adjust and fix the structure with the expanders.
- Add waterproofing to the roof.

You Have Another Option

You'll run into a situation where you have to install a concrete foundation at home. In concrete foundations, you need to consider the weight of the panels, and the structure usually considers a slope of no more than 15°.

Here's what you need to do:

- Make sure that the area where the installation will be made is clean.
- Add the glue to the area.
- Install the hydraulic concrete base on the roof.
- To finish, add the waterproofing to the roof of the house.
- After that, you have to fix it on the roof and put the photovoltaic panels on it, according to the typology and inclination of the roof.

- A flat roof helps the performance of the installation, thanks to the fact that the inclination can be adjusted using the structure.
- The anchoring of the structure will vary depending on the roof to which it is to be anchored. Once this is done, attach and secure the module to the structure using the appropriate fittings.
- Remember that maximum energy will be captured if the tilt is consistent with the geographic height of the installation site. Also, panels are tilted between 20° and 25°, so a reduction in annual performance is acceptable because the sun's declination varies 47° between winter and summer, just as dust or shade can reduce the performance of a solar panel.

Proceed to Connect the Electrical Inverter

In most cases, the interconnection of panels is done in series between modules of the same row or string. However, this depends on the inverter you have.

The interconnection between modules is done through the so-called MC4 connectors. It is important that the stamping is done with appropriate tools, and for the sole use of these connectors. MC4 connectors are used to manufacture mechanical type connectors certified according to quality standards. In addition, it avoids arcing, moisture leakage, and dust in the connections.

You must also make a connection between the series and the inverter to complete the direct current (D.C.) part of the installation. All the energy that each panel collects through solar radiation is conducted as current in the inverter, which will be located at location A. near the load center of the house.

Remember that clamp or twist connectors are required to ensure a reliable connection. So it is not a good solution if you are connecting wires without connectors by taping them together.

There is one very important rule: brute force is never allowed in the connection. The cables must be placed in a way that means that the connection won't be forced when the cable is pulled.

The Inverter

The solar protection of the inverter is very important; that is, it should not be placed outdoors. Otherwise, its performance will be affected when it gets hot. Therefore, it is recommended that you place it in a residential place; if this is impossible put it outside but make sure some elements can protect it.

An inverter converts D.C. power into the A.C. power that is used in all homes. This change is automatic, with a certain magnitude and frequency, so you can have it without worrying about it.

Finally, connect the inverter to the load center as if it were a separate circuit, which completes the A.C. portion of the installation.

Growatt inverters are "plug and play," which means you connect the inverter, turn the power knob, wait for the synchronization time, and the inverter will start working; there is no need to program the device.

Here Are Some Tips and Recommendations

Now that you know how to install solar panels, here are a few more tips so that you can install them without any problem.

The Solar Panel Inclination

Dirt reduces the performance of solar panels. Self-cleaning can be used if the panels are tilted more than 20° and it rains.

Follow the manufacturer's parameters, maintain them frequently and have them ready in perfect condition for when you need to use them for real.

Off-Grid Solar Setup System Components

These are the basic components of off-grid solar panels:

- The solar panels.
- The charge controller.
- The inverters
- The batteries
- The appropriate cables to the panels
- The batteries where the energy will be stored.
- The structures on which the whole system will rest.

Chapter 2
How Much Does the Installation Cost?

Suppose you want to install a solar energy system. Let's talk about the costs and how to calculate what you need.

Cost of Solar Panels for Homes

Photovoltaic installations, at time of writing, cost between $800 and $1000 per square meter. That is to say, for a self-consumption installation with 10 solar panels of 330W, the cost is about 6000 dollars.

This price per square meter of installed solar panels is reduced depending on the size of the system; the larger the installation, the lower the price per square meter; even in large industrial photovoltaic installations, the price ranges between 300 and 400 dollars.

Benefits of Solar Panels for Homes

There are a variety of advantages:

- You will be able to have electricity when the world stops.
- If there is a big blackout, it won't affect you as strongly.
- You will be able to cook recipes in the bunker.
- You will light up at night.
- You will have a system for at least 25 years, with proper maintenance.

Calculating the Cost of Solar Power

It is one of the most common doubts when you want to install solar panels at home. How many solar panels do you need? It depends on several factors that you should take into account when making an approximate preliminary calculation, although it should always be done by professionals and installers later.

Your savings will depend on the energy produced by your solar panels and your home's energy consumption.

Before working out how many solar panels your home will need, you should take into account some important criteria:

- **Your home's energy consumption:** You should look at the kilowatt-hours you consume and, if possible, even differentiate consumption during the hours when solar energy is available (i.e., during the day).
- **Surfaces that can be used to place solar panels on roofs:** are also important. If you do not have enough space on the roof, there is little point in correctly calculating the number of panels you need. Usually, there is no problem with space, but this can limit the installation's power if it is shaded.
- **Roof orientation:** Another key factor is the orientation and tilt of the solar panels. The more space available facing south (southeast, southwest), the better. The farther North you face, the less you'll be able to take advantage of it.

I will show you this in more detail:

- **Total power:** This is calculated by multiplying the number of solar panels by their power. If you need to know how many solar panels are needed for 10kW of power and use 330W panels, you need 30 or 31 panels.
- **Orientation and tilt:** This can affect performance. The

ideal direction is always as far south as possible, with an inclination of about 35°. However, it is also about finding a balance in integration, especially in the home. Sometimes it is not worth breaking aesthetics to get a 2–3% yield.
- **Technology:** If you are looking for maximum performance, you should know that the solar panels you need for your home will always be monocrystalline. Other technologies are cheaper but have lower performance and are less aesthetically pleasing to integrate into the roof of your home.

Energy Consumption at Home

As previously discussed, the ideal is to know your home's energy consumption in kWh/day. But it is important to distinguish between the energy consumption of what is done during the hours of sunshine and what is done during the hours without sunshine.

The reason is logical, and you should take into account that in order to calculate the solar panels you need for your house, you always have to consider when they receive solar radiation.

For this purpose, electricity consumption meters are usually used, and solar panel installers will analyze your home's electricity consumption and properly design your installation.

Calculating an Example So That You Can Apply It

Imagine that you consume hours of sunshine of 8000 kWh/year.

Let's say that the city you are in has an average of 1600 hours of sunshine per year in your geographical area. Peak solar hours are the sum of the annual hours the solar panels are estimated to receive 1000W/m2, i.e., operate at their maximum theoretical output.

You can then approximate the solar panel required in this way:

- House electricity consumption on a sunny day: 8,000kWh/year.
- Peak sunshine hours in your city (map): 1,600HSP
- Total power of solar panels required: (household consumption) / HSP = 5,000Watt.

You will need a solar panel with a total power of 5,000 watts. For example, if you have a 320W solar panel, the math is simple:

(Total solar panel power) / Panel power unit = 15.62 solar panels -> 15 or 16 solar panels will be the final count.

How Many Kwh Does a Solar Panel Produce?

If you want to look at it another way, you can also calculate how many kilowatt-hours the solar panel produces; in this case:

320W * 1,600HSP = 512kwh/year will produce 1 320W solar panel in the area you are in.

Available Space on Your Home Roof

Once you know how many solar panels you'll need, it's time to see how much space you'll need for the P.V. installation and whether your roof has enough space to install it.

One of the tips often given to the installers you work with is that it is best to use medium-sized panels (1.7 x 1.0 m) and black for home solar panel installations.

The dimensions are for ease of installation and adaptation to the roof. The black is purely a matter of aesthetics.

Certainly, if you are putting solar panels on the roof of your house, it better be something pleasing to the eye, not the "fudge" you see every day.

How Many Solar Panels Do I Need for a 200m2 and 100m2 House?

As well as focusing on the surface of your house, you need to be aware of the energy consumption within your house. Consumption is not the same for an average-sized house with constant consumption and a country house with sporadic consumption so it is an important to measure consumption carefully rather than leaving it to guesswork.

Once you have your consumption data, you will have to calculate the number of panels and the space to install them.

Calculating the space required for a solar panel is simple. You have to ask yourself these questions:

- Which solar panel should I install? You have to choose the solar panel you see that gives you good guarantees, quality, and performance. When you do it, look at the cell size of the corresponding panel.
- How much space do I have on the roof? If you have no shade and the roof has a good slope and orientation, you should measure the space you can allocate for the solar panels.

Example of the Surface Area You Might Require

If you continue with the above example, with 16 solar panels, then it would be:

Square surface 1690 mm by 998 mm -> 1,68 m2

The surface area you require on the roof (Panel surface area) x (Number of panels) = 27m2 approx.

This measurement is the raw surface are; then, it will be up to you to see if it should be installed in two rows, three rows, horizontally, or vertically.

When you come to install this system, you need to decide whether you can do it yourself or whether you should go for the option of calling on a professional in order to make sure that you do not make mistakes and then regret it when there is no one to help repair them.

Chapter 3
Basic Electricity Rules

You can't install a solar power system without knowing some of the basics of electricity.

Electricity Basics

Power plants generate electricity by converting primary energy sources (hydropower, thermal energy, solar energy, nuclear energy, wind energy). It is transmitted through the grid to population centers and industries, where it is converted into other forms of energy (secondary energy: light, heat, sound, movement.).

Let's look at the concepts surrounding electricity in more depth:

Electric Current

This is the name given to the displacement of electrons in an electrical conductor. All objects tend to remain electrically neutral, so if two objects come into contact, one with an excess of electrons and the other with a deficit, an exchange of electrons will occur between them until they are electrically equal. The normal direction of current flow is opposite to the movement of electrons, i.e., from + to -.

Electric Circuit

The path that electrons travel. An electrical circuit is similar to a hydraulic circuit in that it can be thought of as the path along which electric current (water) flows from a voltage generator (also called a source) to a consuming device or load. A load consumes energy to generate work: a circuit load can be a lamp, a motor, etc. In a circuit, the current flowing through a conductor depends on the voltage applied to it and the resistance provided by the conductive material. The lower the resistance, the better the current flow.

Electrical Magnitudes

So far, three main electrical units can be defined: voltage, current, and resistance.

Electrical Voltage (U) Or Potential Difference

When there is a difference in the number of electrons between two points, the electrical voltage (or voltage) is called the potential energy (force of attraction) between the 2 points. There is a voltage at the poles of the battery, and the unit of measurement is volts (V).

Electric Current (I)

Electric current is the number or strength of electrons flowing through a conductor, and when a voltage is applied at its ends, it is called current or force. The unit of measurement for current is the ampere (A).

Joule Effect

This is the phenomenon in which electrical energy is converted into thermal energy when an electric current passes through a conductor. This effect occurs in all household appliances (because they heat up when turned on), but some are specifically designed to convert electrical energy into heat (stoves, irons, ovens, thermos flasks) and provide the appropriate resistance. When calculating the heat dissipated, we will use the energy formula in the function of the resistance: $E=Q=RI2t$

The international system's energy unit is the joule, but it is usually expressed in calories when discussing dissipated heat. To convert from joules to calories, it is multiplied by 0.24, so we can transform the above expression to give the result directly in calories:

$Q=0.24RI2t$. Since the joule is such a small unit, it is common to use the kilowatt-hour (kWh) as the unit of energy.

Resistors

Resistance is one of the basic quantities used to measure current and is defined as resistance to current flow. The unit used to measure resistance is the Ohm (Ω), denoted by the letter R.

How is resistance measured?

There are several ways to know the value of a resistor.

The first and simplest is to use a measuring device (ohmmeter or multimeter). To measure with these instruments, simply place the tip on each terminal, and it will automatically give you the value.

All resistors are printed with 4 to 5 colored strips. These strips are crucial because we can color code them, compare them, and know their ohm values.

The third method is more complicated because you have to deal with Ohm's law. And use the corresponding formula to know the resistance value.

Ohm's law: The voltage applied in a circuit is proportional to the strength of the current and inversely proportional to the resistance of the conductor.

Types of Electrical Resistors

Resistors can be classified into three groups:

- Fixed lines, with the value not changing and predetermined by the creator.
- Variable: the value can vary within a predefined range.
- Non-linear, the value varies in a non-linear way depending on various physical quantities such as brightness or temperature.

Power

Electrical power is the rate at which electrical energy is transferred in a circuit per time. More simply, we can think of it as the energy consumed or produced by an element at a given time.

For example, a light bulb can have a power of 12W, thus indicating that it will consume that energy at a given time when it is operating.

In the case of a house, if you install 5kW, that would be the maximum power you can use. You can cook outdoors (2.2kW) and have the washing machine (1.5kW), refrigerator (0.25kW), and heating (1kW) working at the same time because they add up to 4.75kW, but if at that moment you want to use the microwave (0.9 kW) the light will "jump."

A Formula for Calculating Electrical Power

From its definition, it can be deduced that power is calculated by the electrical charge of the potential difference in a finite time.

Generally, one can define the power of an electrical device as the product of the voltage (V) to which it is connected and the strength of the current passing through it (I), resulting in P = V * I, undoubtedly the most famous version of electricity.

Electrical Circuits

A circuit is a set of interconnected electrical elements that make it possible to generate, transmit and use electrical energy to convert it into another type of energy, such as thermal energy (stove), light energy (light bulb), or solar energy. Mechanical (motor). The circuit elements used to achieve this are the following:

- **The generator** is the part of the circuit where electricity is given by maintaining the voltage difference at the ends.
- **Conductor:** The wire through which the electrons driven by the generator go.
- **Electrical resistance:** they are elements of the circuit that oppose the passage of the electrical current.
- **Switch:** it is an element that allows the opening and closing of the flow of current. If it is open, the electrons do not circulate, and if it is closed, it allows the flow of current.

Types of Connections

An electrical connector is a device that allows a secondary device to be connected to a current source. These connections are commonly used in vehicles, appliances, and household appliances.

Connectors for electronic circuits are characterized by having male and female ends that can be connected to form a permanent or temporary connection.

The types of internal electrical connectors are:

- R.J. connector.
- USB connector.
- BNC connector.
- F connector.
- RCA connector.
- Plug.
- Twist-on connector.
- R.F. connector.
- N connector.

Chapter 4
Essential Tools That You Must Always Have Ready

Safety Tools

One of the greatest hazards related to electrical safety is electric shock. The shock causes muscles to contract and, depending on certain factors, such as current and exposure time, can cause cardiac and respiratory arrest. Burns are also a common consequence of this type of accident and the fire risk. People who install and maintain overhead networks, such as light poles and urban wiring, are prone to falls.

Here are the basics you need to keep in mind on the electrical side when installing solar systems:

- Insulated shoes should be worn. These shoes insulate you from the ground, and you should also wear insulating gloves and goggles to protect you in case of a spark.
- Do not wear metal objects when working with the panels. Chains, watches, or rings can cause short circuits or attract arcs. Metal is an excellent conductor of electricity, so touching it can produce a dangerous shock.
- Wear tight-fitting clothing to avoid contact and falls.
- It is best to work without electricity. Most units are segmented so that you can control the current through switches. If necessary, turn off the main power.
- Calculate the amperage before starting work. Use reliable and safe electrical test equipment.

- Avoid working with these systems in damp places or near liquids.
- Analyze circuits and connections. Investigate the composition and characteristics of circuits before you begin work to assess the hazards and develop safety standards appropriate for the type of circuit you are using.
- Whenever possible, work with only one hand. If you receive an electric shock, the electricity passes from one hand to the other and through the heart.
- When you install electrical equipment, you must leave free space so that it can be operated without difficulty in the future. All parts of the circuit must be available at all times.
- The fuse must be well protected to prevent external components from entering the area.
- Use your tools responsibly. Fortunately, many ancillary materials are available today, but sometimes we use tools for non-design purposes. Use a complete set of tools without any risk.

In addition, you must have:

A Safety Helmet

When working with panels, you should pay special attention to electrical safety and hygiene rules. Electricians must wear a type A helmet permanently as it protects from the risk of shocks, impacts, and splashes of pyrogenic material. Safety helmets are materials resistant to fire, solvents, impacts, and abrasion with low electrical conductivity. The most commonly used materials are high-strength laminated plastic and resin-impregnated fiberglass.

Safety Glasses

These are glasses with side protection, transparent on the inside and dark on the outside.

Safety Clothing

Clothes that accumulate static electricity or that are especially flammable include synthetic fibres like polyester. Proper protective safety clothing that engineers working with electricity use is not the easiest or most comfortable option for a prepper.

100% cotton and wool will burn, but it does not have the same risk as synthetic fabrics which can melt and stick to the skin, so there is obviously always still a risk when working in a situation where sparks can be generated, but wearing cotton or wool reduces the risk of burning.

Heavyweight cotton, like denim, or a mix of cotton and wool, can provide more protecti

100% cotton and wool"Lightweight cottons and wools sometimes burned, but without the melting and sticking of synthetics. Heavyweight cottons, wools, and blends of the two did not burn.... Heavyweight means that a material weighs at least 11 ounces per yard, like the fabric in a denim jacket."

othat have the risk of sparks due to accumulation of static electricity, the use of polyester and others are made conductive made of electrostatically discharged synthetic fibers. There is specific safety clothing available that is It is wise to explore whether you can buy clothing yourself on eleIn electrical safety and hygiene standards, safety clothing, especially work clothes, must be made of cotton for low voltage tasks, but for high voltage tasks or

Safety Belt

When working at heights, you must use a full-body harness and other elements to arrest a fall in case of an accident.

Gloves

Rubber: They have specific uses in electrical tasks. Reinforced leather gloves are worn over them to prevent abrasion or punctures. They are used in low voltage energized circuits.

Safety Shoes

Insulated safety shoes without studs, eyelets, or metal parts, but with insulation covering the toe cap, should be worn when working. Insulation is achieved through the use of rubber compounds. And you should always wear dry shoes.

Face Protection

You should wear a respirator when drilling concrete or other materials for wiring in walls or floors when working. You know the hazards for electricians.

Here are the main electrical health and safety rules you should follow, whether working from home or electrical circuits.

Power Tools

These are the tools you need to do the installation:

- Current clamp
- Solar irradiance meter
- Insulation multimeter
- Digital multimeter
- Earth resistance clamp

- Battery Analyzer
- Infrared thermometer
- Long nose pliers

Writing Tools

These are the tools you need to use for writing:

- A notebook to take notes about the maintenance of the panels.
- A notebook where you write down the number of panels you need.
- Information on parts replacement, repairs, and a log of your solar system.

Solar Racking Tools

There are a variety of solar racks to consider for your solar racking system.

Solar mounting systems or solar panel mounts support solar panels on any surface, usually on a roof or ground. Roof mounts are usually less expensive because they can use the structural support of an existing roof. However, ground-mounted systems are easier to access and maintain and do not include the safety concerns of working on a roof.

Most supports consist of a frame attached to the roof truss and footings on roofs. If roof penetration systems are not required, such as clay tile roofs, metal roofs, or flat pool roofs, racking systems can be free-standing and ballasted.

Ground-mounted racking systems are metal frames attached to concrete slabs or mounted on poles to facilitate space underneath, such as in areas with heavy snow, or dual-purpose systems, such as agricultural photovoltaics that combine farming with solar panels.

Most of the racking components for floor and roof systems are high-quality aluminum and stainless steel. An important consideration is the strength of the racking, which must withstand snow and high winds in many areas. A rugged racking system can withstand snow loads up to 90 pounds per square foot and winds of 190 mph.

Ground-mounted systems using frames are similar to roof systems, except that the frame is placed on a concrete slab, which requires additional excavation, foundation placement, and cement steps. Many ground-mounted frames can be manually adjusted to maximize sun exposure.

The pole-mounted system uses a pole secured in a concrete-filled hole approximately half the length. Pole-mounted systems can be equipped with automatic solar trackers or manual adjustments. These systems can use multiple studs to support larger arrays of panels with a smaller footprint than when the frame is placed directly on a concrete slab.

Battery and Maintenance Tools

These are tools you need when you go to work to do battery maintenance:

- Rubber gloves
- Protective goggles
- Ammeter clamps
- Protective lacquer
- Metallic brush
- Funnel and distilled water

Chapter 5
How to Calculate Your Electricity Needs

Calculating energy needs and reducing and harnessing everything will help you get the best performance from your panels.

How to Calculate Energy Needs

In the case of stand-alone (off-grid) solar photovoltaic installations, the correct dimensioning is fundamental to guarantee your energy needs and limit the economic cost of the installation.

Let's imagine a house that is used by a family of four at weekends. Currently, it has no electricity. These are the calculations that that family might make:

Step 1: Calculate the Estimated Consumption

You assemble the basic equipment needed for the example that will consume energy:

- **Light bulbs:** 4 units x 4 hours x 60 Watts (100%) = 960 Wh
- **Television:** 1 unit x 3 h x 70 W (100%) = 210 Wh
- **Laptop:** 2.5 h x 60 W (100%) = 150 Wh
- **Refrigerator:** 24 h x 200 W (50%) = 2400 Wh
- **Microwave:** 0,5 h x 800 W (100%) = 400 Wh

In this section, you will have to estimate the consumption for your specific situation. Here you can estimate the necessary consumption of other devices, such as the need for self-consumption to partially satisfy the needs of devices connected to the grid or devices intended to provide recharging points for bicycles, motorcycles, electric vehicles, battery charging, etc.

Then, if you add up the different partial consumptions, you get the total estimated consumption for the house:

Total consumption estimated per day CDE = 4120 Wh/day.

You apply 75% of the performance you install to calculate the total energy to supply the demand.

The total energy needed: Ten = Cde / 0,75 = 5493 Wh/day

The other thing to look at is the available solar radiation.

To obtain the incident solar radiation, you can use a table with current estimates. A good source for these estimates is the PVGIS application (Photovoltaic Geographic Information System—European Commission, Joint Research Center), an online platform that provides easy and quick access to insolation data across Europe.

Assuming your installation is in a city X, using the PVGIS application, you could obtain the following values:

- **Latitude:** 37°10'38" North
- **Longitude:** 3°35'54" West
- **Nominal power of the solar photovoltaic system:** 1kWp
- **Module inclination:** 35deg.
- **Orientation of the modules:** 0deg.

Where:

- **Ed:** is the average daily electrical energy production of the system kWh.

- **EM:** average monthly production of electrical energy of the system kWh.
- **Hd:** the average daily sum of global irradiation per square meter received by the system modules (kWh/m2).

The most unfavorable month for irradiation is December, with 4.27 kWh m2/day. Therefore, you adjust the installation size according to the most unfavorable monthly insolation conditions to ensure that the demand is covered throughout the year.

Once you know the incident solar radiation, divide it by the incident solar radiation you use to calibrate the module (1 kW/m2), and you get the peak sunshine hours (HSP). For practical purposes, this value will not change in our example, but we will use the concept of HSP (Hours of Peak Sunshine), which is the amount of full sunlight the sun must receive at 1000 W/m2 to get a day because, in reality, the intensity of the sun varies throughout the day.

HSP = which is the solar radiation tables / 1kW/m2 = 4,27 HSP

Step 2: Calculate the Required Solar Panels

Determine the number of modules (or solar panels) based on the most unfavorable radiation conditions. For this calculation, assume that you choose a 180 W module. This data is given in the technical characteristics of the selected modules according to each model and manufacturer.

For daily use installations, you will use the formula:

Number of modules = (energy required) / (HSP * operating efficiency * peak module power).

The operating efficiency considers losses due to possible scaling and degradation of the P.V. panels (typically 0.7 - 0.8).

The number of modules for daily use installation:

Nmd = (5493) / (4.27 * 0.8 * 180) = 8.9 can be rounded to 9 modules.

For weekend installations, you use this formula:

Number of modules = (3 * energy required) / (HSP * work output * 7 * peak power of the module)

Number of modules for you to install for weekend use:

Nmfd= (3 * 5493) / (4.27 * 0.8 * 7 * 180) = 3.8 Redoing 4 modules.

Since our example case is for a weekend house, you require four modules of 180 W each. Remember, the consumption demand established here is very basic; if the first section introduces a higher consumption, it will result in more dishes.

With 180 Watts Peak (Wp) of modules selected, we get a total solar installation of 720 Wp (4 x 180 Wp).

Considering that the modules operate at 12V, if you want an installation at 24V, you can associate the two sets of plates in series and then connect the two sets of two plates in series and parallel. The operating voltage will depend on the battery system we choose.

Step 3: Battery Capacity

To design the capacity of the accumulators, we first have to establish the required autonomy on unfavorable days without sunlight due to cloudy conditions.

In the current situation, the maximum required autonomy can be established in 3 days (Friday, Saturday, and Sunday) at the weekend. In rural residential electrification, the daily power supply can be established between 4–6 days, considering that if we have backup generators, this value can be lower.

Battery capacity is the energy required per day of autonomy/ per voltage per depth of discharge of the battery. The depth of discharge depends on the type of battery selected. These values range from 0.5 to 0.8. You can consult these values in the technical characteristics of each model and manufacturer. In our example, we will choose a battery that allows up to 60% (0.6) of discharge.

It is the capacity to accumulate = 5493 * 3) / (24 * 0,6) = 1144,38 Ah (c100) The value c100 indicates that a 100-hour charging cycle will provide the battery capacity, which is the charging frequency typically established in rural electrification.

The choice of an accumulative system requires different controls to be durable and have optimal performance. The energy storage system requires a minimum charging intensity to ensure that the batteries are charged properly and to avoid a shorter than expected battery life.

This section is intended to serve as a basic example to calculate the necessary parameters to perform an installation, but once we know the capacity required for the installation, I recommend contacting an expert for more details or information on the technical characteristics of the specific system the equipment.

Step 4: Selection of Regulators and Converters

All that remains is to choose a charge controller and a D.C. to A.C. converter so that your home can have 220 V A.C. for any type of device or appliance.

The charge regulator is determined by the maximum operating current and voltage designed and installed.

The power of the DC/AC converter should be selected based on the sum of all the rated powers of the consuming devices multiplied by the simultaneous utilization factor of these devices (Typically, the values range from 0.5 to 0.7). In this example, the total estimated power is 1360 W

Converter power = 1360 * 0.7 = 952 W

Therefore, as long as you are using the device as originally intended, for our example, a 1000 W converter is sufficient. You can always set a higher power if other devices with higher power consumption are used in time.

Let's now look at an example for sizing an off-grid solar P.V. system. The size of this equipment minimizes the consumption to be made to obtain affordable equipment, but this means giving up some elements of higher consumption and power. Therefore, in this section, washing machines, ovens, etc., are not considered, as in the case of a house with very basic electrical installations for a second family.

Although the first step in saving energy with renewable energies is to reduce consumption, making a basic installation is not always possible, as suggested in the example. For other types of homes or buildings, there are other options for saving energy with solar photovoltaic installations. One option is to make self-consumption solar installations connected to the grid. In these cases, the installation is complimentary, allowing you to save but at the same time stay connected to the grid.

How to Reduce Electrical Needs

Reducing energy consumption is undoubtedly the best way to reduce overall energy consumption. Even with the most efficient energy production facilities, the cheapest energy is still the energy not consumed. So how do you reduce energy consumption when isolated at home with every resource worth gold? These are the steps that you should follow:

First of all, less consumption means more output for each element. However, the financial aspect is by no means the only important factor to consider in reducing energy consumption.

When you are cooped up at home, with a single, unique system, you have to use energy responsibly, where electricity is for living, food preparation, and everything vital while cooped up with your family. Keep in mind that you will be cooped up all day long.

Use Less and Better Electricity

To reduce energy consumption, you need to reduce the cost of various energy sources, such as kilowatt-hours, and avoid heating. For example, when calculating electricity consumption, we consider the number of homes and establishments heated at a higher power than necessary, which means useless electricity consumption (or any other type), with the consequent cost of this necessity.

Small Daily Actions to Reduce Energy Consumption

In this sense, there are no secrets: turn off the lights when you leave the room, choose energy-saving LED bulbs, invest in energy-saving appliances (check the energy card of your appliance), avoid turning on the air conditioner, use the oven sparingly, or turn off appliances (computers, routers, etc. console, T.V., etc.).

Everyone knows that these small daily actions are very effective in saving energy. However, we regularly forget to implement them.

Choose Renewable Energy Instead of Fossil Fuel Energy

Certain types of energy, such as solar or wind power, exist in abundance. They are renewable and more efficient sources of energy. Therefore, the best thing you can do is to make better use of renewable energy. Solar photovoltaic is not only the most abundant system but also the most recommended for producing domestic hot water and heating. The consumption of this renewable energy is free, just like wind and biomass.

Therefore, your choices will depend on your consumption and the number of systems you install.

Invest in an Efficient Thermostat to Optimize Heating

Heating can account for more than half of a home's energy consumption. It is, therefore, an important factor when you consider energy savings. That is why our best ally would be a room thermostat with a counter, connected and intelligent to optimize the operation of the boiler and radiator.

A good thermostat allows you to program the ideal temperature for each house room at any day. A connected thermostat goes one step further because it considers different parameters, such as weather conditions or living habits. It is the ideal solution to reduce the energy consumption associated with heating.

The Number of Solar Panels You Require

The number of solar panels you require at home will depend on how much you consume and how you consume them. Here is a brief overview of the indicative table according to the level of consumption.

- Up to 2000 kWh: 2 to 4 panels.
- From 2000 to 5000 kWh: 5 to 7 panels.
- More than 5000 kWh: from 7 panels.

Chapter 6
Building and Installing an Off-Grid Solar System

I want to go deeper into the way to install your solar panel system, the wiring, and how to sort out the area where you place the batteries.

The Best Way to Install Solar Panels

Installations that are not connected to the grid are called off-grid photovoltaic installations. These installations are designed for when you need extra electricity or are isolated.

Therefore, solar panel systems must have photovoltaic batteries to use the energy when they are not in production, so that at night you will have light, and during the day, this battery will be topped by the UV rays.

Once you have completed the administrative process that you have to follow to install your solar system at home, you can follow this process for the installation.

How to Wire Your Off-Grid PV Solar System

The use of photovoltaic systems is an important pillar of renewable energy, helping to ensure that you have electricity when disaster strikes. Photovoltaics are used for various applications, from large power plants to domestic applications.

What Are the Special Requirements for Conductors and Cables in Photovoltaic Systems?

Modern PV systems produce much higher yields than a few years ago. The higher system voltages require stronger cables and components. PV systems typically operate for 20 years or more. Therefore, the cables must be durable, high-quality materials, as they are constantly exposed to wind and various weather conditions. Therefore, their components must be especially resistant to weathering, heat, wind, and frost for reliable performance. They must also be rodent-proof.

To exploit the full potential of the energy supply, the industry is using new technologies and applications, such as organic photovoltaics (OPV). The PV industry's requirements for cabling solutions range from robust, large-diameter subway cables for power plants to filigree connection technology for flexible OPV modules.

Why Are LAPP Products Suitable for PV Systems?

You are constantly developing products and solutions: the highest possible durability, robustness, or extreme flexibility are features you always have in mind. Thanks to the special jacket material, LAPP solar cables are not only flame-retardant and halogen-free but also resistant to ozone and UV rays. Our solar cables and connectors can withstand even extreme weather conditions such as heat, wind, and frost, making them ideal for use in your outdoor installations.

Building Your Battery House

First, all batteries involved must be identical (twins), and all must have the same charge level. Secondly, it is important to use short cables of equal length and adequate cross-section.

The parallel connection of two identical cells can double the capacity of a single cell while maintaining the same nominal voltage.

In this example, two 12V 200Ah batteries are connected in parallel, so you get a voltage of 12V (volts) and a total capacity of 400Ah (amp hours). The capacity identifies the maximum amount of charge that can be stored. The higher the capacity, the greater the amount of charge stored. It is measured in amp-hours.

In this case, with a capacity of 400 Ah, the battery pack can theoretically deliver 400 A continuously for one hour, 200 A for two hours, 100 A for four hours, and so on. The lower the maximum current drawn by the battery, the longer it will last.

By connecting two identical cells in series, you can get twice the rated voltage from a single cell while maintaining the same capacity.

In this example, with two 12V 200Ah batteries in series, you will achieve an output voltage of 24V (volts) and a constant capacity of 200Ah (ampere-hours).

In stand-alone wind and PV systems, the higher the voltage used to charge the battery, the lower the energy loss in the cable. For example, a 24 V system is better than a 12 V system.

Combining parallel with series will double the rated voltage and capacity.

We will have two 24V 200Ah blocks connected in parallel following this example, resulting in 24V 400Ah.

During connection, it is important to pay attention to polarity and use cables of sufficient cross-section and as short as possible. The shorter the connection length, the lower the resistance created in the cable when current flows and, therefore, the less energy is lost in the cable.

A large and efficient storage system is essential when designing a stand-alone PV system. I recommend using a high-quality, high-efficiency charge controller to ensure proper battery charging. The charge controllers you have selected are designed to ensure an optimal charging process for each type of battery, including lithium iron phosphate and lithium iron phosphate, using MPPT technology to harness all the power generated by the PV panels.

If you want to convert the battery's DC into a household AC, you must purchase a sine wave inverter to power any household appliance. There are two types: modified sine wave inverters (suitable for resistive and capacitive loads; possible noise with inductive loads) and pure sine wave inverters (suitable for all loads).

As for the location, you have to put it at home, in an area where there is no humidity, no high temperatures, and it does not get sun or wet. You have to take care of them to avoid accidents and ensure that the system works.

I recommend that you design a structure a little high off the ground so that you do not trip over it and manipulate it without any problem. Besides, you will be able to keep it safe. If you can, create a structure where the goals are open to the air but protected.

Chapter 7
Maintenance of an Off-Grid Solar Power System

When you are isolated from everything when you are dependent on the PV system, you have to learn how to maintain it on your own; let's see how you can take care of it.

Types of PV Maintenance

There are two types of maintenance for PV systems: corrective and preventive. The first is when the system fails. In most cases, it manifests itself as loss of power or no power production for various reasons, in which case the system must be corrected (replacement of components, parts, etc.). It is called corrective maintenance. It often happens because you do not do preventive maintenance designed to check the condition of the system components to verify that everything is in order.

The frequency of this maintenance depends largely on each user. However, I recommend preventive maintenance of the PV system every 6 months or at least once a year. In addition, you must be very knowledgeable about the installation.

WARNING: Preventive maintenance will not prevent your system from failing, which can be a factor if you have poor installation or low-quality panels and inverters. If you want to avoid major problems, buy the best quality ones later.

Factors to Consider When Installing Bunker Panels

What do you do with preventive maintenance?

The main activity in maintenance is to check for environmental factors that affect the system's operation: dust, water, moisture, humidity, dirt, etc.

The work that is done in a good maintenance service is as follows:

- Observe all power cables, conduits, and cable ties.
- Check and clean junction boxes and protective devices.
- Clean the solar panels.
- Re-tighten inverter connections and clean the interior.

As you can see, the inverter cooling channels may be clogged with a thick layer of dust. In this case, preventive maintenance can remove all layers of dust and allow the inverter fans to operate in optimal conditions. If this is not done, the dust can block the cooling system, reduce inverter efficiency or even damage the inverter.

For your solar system to last more than 25 years, it is necessary to perform the preventive maintenance recommended by the brands on time. Remember, it's not just about cleaning surfaces; every part of the system is examined. The maintenance costs are nothing compared to the operational problems that can arise if you do not do the maintenance, not to mention if it is damaged when you are locked in the bunker and have no one to repair it or the necessary spare parts. Remember that the recommended maintenance frequency is every 6 months, and, ideally a prepper would expect a system to last more than 25 years.

Maintenance of Different PV Components

Solar photovoltaic panels require little maintenance but making sure that the surfaces you install them on are perfect will extend their useful life and improve their performance.

That is why I am now dedicated to giving you information on how to do the maintenance and cleaning of solar panels step by step.

Dust, dirt, pollen, bird droppings, and leaves can affect energy production, so regular cleaning can do wonders and increase the efficiency of your appliances.

How, When, and How Often Do You Start Cleaning Solar Panels?

Remember not to turn off the system

If necessary, the device can only be turned off completely if you are going to do any repairs.

You don't need to get too complicated when cleaning solar panels or invest in complex solutions. Simple mild soap and water are enough to remove accumulated dirt and debris.

If the dirt is more persistent, a scraper or sponge made of soft material and an extendable handle is required to reach hard-to-reach places and avoid damaging the panel surface.

It is best to consult the manufacturer's instructions or operator's manual for the best information on what cleaning products or materials should or should not be used.

However, despite the simplicity of the task, it is best to clean with great care to detect damage or check the condition of the PV system.

Use a hose whenever possible, and direct the water towards the top. While it's not a problem if a little water touches the back of the panel, try not to do it directly. Also, you always have to put safety first before cleaning or starting work and have all the necessary equipment (if needed) ready to climb on roofs and hard-to-reach places to keep yourself safe (seat belts, ladders, footwear, etc.).

When and How Often to Clean Solar Panels

Late spring and late fall are good times to clean solar panels. Choose a time of day when the weather is cool and mild. Think of it like a shot glass: panels with cold water can increase the likelihood of cracking due to sudden temperature changes. Also, if the sun is in full swing and the surface receives a lot of radiation, there will be a lot of heat, so the water you use can evaporate quickly and leave dirty marks.

It's especially good to clean solar panels first thing in the morning, as the spray from an overnight installation will help remove accumulated dirt, which means you'll need to use less water and less energy to clean. The plate requires good maintenance.

Cloudy days or evenings are also ideal times to clean your solar panels. Suppose you live in a particularly dry area, a rural area, or an area with high air pollution, including an industrial area or one affected by nearby buildings. In that case, you may need to maintain your installation more frequently. But this should be no more frequent than every 3 to 4 months.

Generally, it is recommended that you clean about 3 or 4 times a year, but in areas with more exposure to the solar panels, such as rural areas or near the sea, more thorough and frequent maintenance is recommended, as well as more frequent checks of the condition of the photovoltaic and solar installations to predict possible system failures. This should also be considered if the panel's surface is more horizontal and angled.

Visual Inspection of Photovoltaic Panels

- Check the structure of the support panel. The panels are usually aluminum, with stainless steel components that do not require anti-corrosion maintenance. But it is important to observe the cover for anomalies, such as cracks or stagnation on the board's surface, and if it is well fastened.
- Check electronic components. Monitor DC panels, AC power, inverters, monitoring systems, etc.
- Check the storage system. In isolated installations, i.e., installations with storage systems (solar cells), you should consider keeping the top surfaces of the cells and terminals and the connection terminals clean. Also, on a more technical level, when the battery needs maintenance:

- If necessary, recharge the battery as indicated in the factory instructions to prolong its life and ensure good performance.
- Use a hydrometer to check the battery's state of charge and capacity by measuring the electrolyte density. You may not have this equipment; if you cannot take this measurement, try to call a professional capable of handling the task.
- Periodically equalize the battery to regulate its storage capacity, improve efficiency and prolong its life.

Other Solar Panel Maintenance Tasks

It may be that your installation is in an inaccessible location or at an inaccessible height, and you consider using a pressurized hose system to clean your solar panels. It is not advisable to do this. The force of the pressurized water can damage the surface of the panel, so it does more harm than good. Also, don't attempt to enter a high-rise installation without proper safety equipment.

Working on a roof is high risk and increases when water enters the equation, making any operation more slippery and dangerous. It is best to hire a professional company or a qualified professional for your safety.

If your panels are covered with accumulated dust or dirt, your stand-alone installation may be less efficient. Panels close to the sea are also affected by saline environments.

Tips for Cleaning and Maintaining Solar Power Over Time

Keep these general tips in mind:

- Wash the module with water, only the upper part.
- If you need to use something other than water, you can use a mild detergent. Do not use harsh, abrasive cleaners.

- Windshield wipers can be used to make cleaning easier. Again, do not use clothes that can damage the panels.
- Clean the panels early or late when they are cool to avoid a thermal shock.
- You can dry them with a soft cloth or windshield wiper.
- Avoid shade, caused by trees and other vegetation, that may cast shade on the solar panels. It reduces their performance.
- Seek expert advice on solar energy issues: when you decide to install a solar energy system, look for someone trained to ensure a good installation. This way, they will do the right research for your specific case, and you will have professionals by your side who can always advise you in the future if necessary.

In terms of frequency

The weather and the appearance of the solar panels will be key indicators to consider when thinking about how often to clean and maintain them. The frequency can be 1 to 2 times a year, but it always depends on the conditions they face and the region of the country where the installation is located.

The Step-by-Step Check of the Solar Kit Spaces

Solar inverters are distinct elements and must be checked step by step. It is important to note that they should not be placed where they will not be exposed to very high temperatures (above 35°).

- Check the wiring annually, as it will degrade over time.
- Verify that the voltage and current are correct.
- Check the indications and warnings of operating errors. For this, you have to refer to the specific model's manuals.

Solar Cells

If your solar panel kit has batteries, you must go through a series of care and maintenance routines.

- Perform exterior cleaning and damage inspections at least once a year.
- In addition to cleaning, check for potentially damaged areas, etc., verify that connections are intact and secure.
- Check that the water level is correct four to five times a year.

Regular maintenance of the entire installation is essential in order to ensure the longevity of the photovoltaic kit; not only that, but you will also guarantee optimal performance, which means that the solar panel kit will generate more energy.

Chapter 8
Solar Panel Elements

It is important to understand the differences between solar panels and the functions that each type has, so that you can know how to make the best use of them.

Solar Charge Controllers

A charge controller is an electronic device whose function is to control the state of charge of a battery to ensure optimal charging, thus extending the battery's life.

Installed between the photovoltaic field and the battery, the solar charge conditioner is responsible for controlling the flow of energy between the two elements. This control of the energy transfer arises due to the control of the current (I) and voltage (V) parameters throughout each charging phase.

Charge Regulator Characteristics

- The regulator is configured for the connected battery and will apply the appropriate algorithm to maximize battery life.
- It also protects the battery from possible overcharges and surges, compensating for the higher voltage of the PV field so that the battery is not damaged depending on the state of charge at any given time.
- Depending on the manufacturer, the regulator can be complemented with an external display or a communication device, or if it is of the 3-in-1 type, it can be integrated into the inverter itself.

Types of Charge Controllers

The charge controller directs and controls the energy flow between the battery and the solar panel. There are 2 types: PWM charge controller and MPPT.

PWM Charge Controller

It pulses, cutting off the power flow between the panels and batteries only when fully charged. To work properly, it must have the same nominal voltage on the solar panel and battery, i.e., if we have a 12V battery, we can only charge it with the 12V panel. Solar panels work with the battery's voltage during the charging phase, an operating point that is not under the maximum intensity that the panel can provide, so they are produced without all the solar energy and are more economical.

If you require more charging power, you can connect more 12V panels in parallel in order to not increase the voltage.

With a PWM charge regulator, the module works with the battery charging voltage, which results in power loss. Once the battery reaches the indicated voltage, it starts to block the contact between the module and the battery to avoid overcharging, this is called the absorption phase, but this leads to a drop in energy efficiency, which is a big obstacle. Its strengths lie in its price and ease of transportation due to being lightweight.

MPPT Charge Controller

They are also called maximizers because their operation uses the maximum output of the solar panels to charge the batteries. In addition to cutting the current to the battery while charging, this type of regulator also obtains the highest output from the panel, allowing it to operate at maximum. It internally adjusts the voltage, which is always higher than the battery, to the required voltage, with high conversion efficiency, gaining strength and saving total energy output. It is the best option to get the most out of your solar panel, where the extra cost more than compensates for its superior production capacity. While they can work with panels and cells at the same nominal voltage, they work more efficiently if we increase the PV field voltage.

Unlike PWM, MPPT regulators include Maximum Power Point Tracking (hence the acronym) and a DC-DC transformer (which converts HVDC to HVDC while charging the battery). This regulator works with the module at the most appropriate voltage at any given time to extract maximum power or limit it to the sinking or floating phase (phase in which the regulator must keep the battery charged and avoid overcharging and discharging).

MPPT regulators are more useful because there are more PV modules connected to them than PWM due to voltage incompatibility and more of these outputs. However, if we work in low-power PV, we can also use a PWM charge controller.

How to Choose a Charge Controller?

The PWM regulator provides several numbers: operating voltage (12, 24, 48) and current (10, 20, 30... amps). We can only use panels at that voltage to charge a 12V system. We must never exceed the current and power rating indicated by the regulators that produce our panels in parallel.

The MPPT regulator provides 3 numbers: the operating voltage at the battery, the maximum operating voltage at the panel, and the battery charge strength. In no case should we exceed the operating voltage at the panel, which will allow us to arrange its connections so that it is always higher than the battery voltage and look at the VOC numbers on the panel datasheet. On the other hand, the battery charge strength will tell us the total power of the panel that the regulator can handle. If it is 200W at 12V, it will be 400W at 24V. It can exceed the installed power figures, limiting the regulator to the maximum power it can charge.

Finally, you must remember that some panels will charge the battery correctly only if you use MPPT regulators and connect them in pairs or groups of three. In some cases, they will not produce enough voltage to power the battery fully charged, thus damaging the battery prematurely.

My advice is to follow the manufacturer's instructions, and if you have any doubts, ask and consult, as many failures in solar systems are caused by bad connections between the panel and the charge controller.

Solar Battery Bank

The battery pack of the photovoltaic system or accumulator is responsible for storing the energy so that it can be supplied at that precise moment independent of the electrical production of the photovoltaic generator as a reserve for cloudy days or nights. They consist mainly of two electrodes immersed in an electrolyte, causing chemical reactions when charging and discharging.

Unit of Measurement of Energy Capacity

The capacity of a battery is determined by the amount of energy that can be extracted from it before it is completely discharged (when fully charged). Its capacity is measured in amperes per hour (Ah) for a given discharge time.

The battery's capacity can be defined; for example, if its capacity is given as C20, this represents the amount of energy extracted from the battery in 20 hours at a temperature of 20°C until the voltage enters each terminal block Glass 1.8V. If it is discharged quickly, the battery capacity will decrease; conversely, if the discharge time increases and becomes slower, the battery capacity will be higher. So, if we have a battery with C100=250Ah, the battery can provide 250A for 100 hours.

Manufacturers usually provide information about the battery capacity at different discharge times. The C100 capacity is often used for calculations.

Depth of Discharge

This is the ratio (expressed as a percentage) between the charge obtained from the battery and the rated capacity. For example, if a 250 Ah battery is subjected to a 100 Ah discharge, it is subjected to a 40% discharge of the entire battery. The battery's service life is determined by the number of charges and discharge cycles it can withstand at a given depth of discharge. It is also proportional to the usual depth of discharge. For example, in a single cell with a lifetime of 180 cycles and a depth of discharge of 80%, the discharge is reduced by 30%, which means that the battery's lifetime can be increased to 1000 cycles.

Life Expectancy

Life expectancy is generally considered to be measured in years. However, manufacturers specify that the expected life expectancy of a battery is based on its number of cycles. Batteries lose capacity over time, and when they lose 20% of their original capacity, they are considered to have reached the end of their useful life, even if they are still usable. It should be taken into account when designing the system from the beginning.

Depth of discharge DOD (depth of discharge) refers to the percentage of capacity used. Battery life (number of charges and discharge cycles) versus DOD (battery capacity percentage). A shallow cycle battery, below 25% DOD, is estimated to have a life of at least 4000 cycles. If the same battery were used but cycled at 80% DOD, its lifetime would be reduced to 1500 cycles (considering that 1 cycle equals 1 day).

Solar Inverters

Solar inverters, also known as photovoltaic inverters, convert the direct current generated by solar panels into alternating current, which can be used in the home, stored in solar cells, or injected into the grid.

However, when we talk about these systems, we usually think only of solar panels installed on roofs or other spaces, but solar systems also have other elements necessary to ensure their operation, among which inverters stand out.

Also known as a photovoltaic inverter, this component can convert the energy produced by the solar panels for use in the home, studio, or place of refuge.

Hence the name inverter is responsible for converting the DC output of the solar panel to AC, which is used by all commercial appliances.

How the Solar Inverter Works

Solar panels are responsible for receiving solar radiation and converting it into electricity through the photovoltaic process. Solar panels consist of positive and negative layers of crystalline silicon or gallium arsenide semiconductors connected by junctions. These layers absorb sunlight and send its energy to the photovoltaic cells.

However, this current is direct current (DC) and variable, so a solar inverter is required to convert it to alternating current (AC). In Spain, the standard used is 230 V, so the inverter is responsible for supplying this voltage so that the various appliances and electronic devices can function optimally.

In addition to being responsible for converting direct current into alternating current, solar inverters have other important functions:

- **Improve the performance of the installation:** the inverter is responsible for using all the electrical energy generated by the solar panel and, at the same time, stabilizing the electrical wave to ensure a stable power supply within the specified range, avoiding damage to household appliances and other electrical and electronic equipment.

- **Control and protection:** The inverter continuously monitors the system and analyzes energy performance. It detects any event and warns of possible errors; it is also responsible for dissipating the heat generated.

How are solar inverters classified?

Solar inverters are classified according to whether they are grid-connected or off-grid photovoltaic systems.

Grid-Connected Inverter

Solar inverters that connect photovoltaic installations, either for use or to sell energy, need to be synchronized with the traditional grid. They are called grid-connected inverters, and 3 types can be found in the current market:

- **String-connected inverter:** This is the type of inverter most commonly used in self-use systems. The solar panels are connected in series, and the energy they produce goes directly to the inverter. When the panels are not affected by shading, they are highly efficient, and optimizers can be installed to increase efficiency.
- **String-to-grid inverter with optimizer:** The optimizer is placed next to or integrated into each solar panel. The energy obtained from each panel is sent to the string inverter, increasing the system's efficiency. In addition, the performance of each solar panel can be monitored individually.
- **Distributed inverters or microinverters:** These are placed individually on each solar panel to minimize the negative effects of shading, as the PV system will still generate power, although some panels may have a lower yield due to shading. They also allow the efficiency of each panel to be monitored individually.

The Inverter for Off-Grid PV Systems

This inverter is used in photovoltaic installations and is not connected to the general grid, so batteries must be present. They are mainly used in photovoltaic power generation systems in remote areas or ships.

They are responsible for converting battery current to 220V, providing efficient power for appliances and other electrical or electronic devices. They are programmed to cut power when the voltage is too low. They also allow the installation of a charger that is activated to charge the battery when the voltage is too low.

Hybrid Inverter

They are designed to collect energy from the general grid and solar cells. They are used in places where power supply is difficult.

Conductors and Connectors

The MC4 allows modules to be connected so that panels (or strings) can be easily constructed by simply pushing the connectors of adjacent panels together by hand. In the event of disconnection, they must be disconnected using a special tool called an MC4 connector wrench; to ensure that they are not accidentally disconnected by pulling on the cable. All solar panels produced since 2011 feature MC4 and are universal in the solar market.

Design and Structure

The MC4 system consists of a plug and socket design. The male and female connectors are placed inside a plastic housing. For proper sealing, the MC4 requires the correct cable diameter. Typically, double insulation and UV protection (most cables will deteriorate if used outdoors without sunlight).

Typically, a pair of MC4 consists of a total of 10 pieces, 5 pieces per connector, as follows:

Important notes

Even in low voltage systems (12-48 V), it is very important not to connect or disconnect under load. Because arcing can occur and severely damage the materials in contact, resulting in high resistance and overheating, this is partly due to direct current (DC), whereas if used for alternating current (AC), the arc will self-extinguish at the point where the voltage is at zero.

Plug connectors that are not manufactured by Multi-Contact (MC) but that mate with the manufacturer's Multi-Contact (MC) products and are sometimes described by their manufacturer as "MC Compatible" are not eligible. Long-term safe and stable electrical connections, for safety reasons, it is not recommended to connect them. Therefore, MC is not responsible if these non-approved connectors are mated with MC components.

To avoid arcing when the MC4 is disconnected, a breaker should be installed near the connection to open the circuit without the risk of arcing. MC4 connectors are rated for 30 A and are designed for 12 and 10 AWG (4 and 6 mm2) conductor sizes.

To properly install the MC4 connector, 2 basic tools are required. The first essential tool is the MC4 punch down clamp, which is used to clamp the pins; it is very important because they work with the correct pressure to clamp the pins to the cable.

Using common jigs or simply bending the retaining pins by hand is risky because of the risk of arcing due to loose pins.

The second tool is an MC4 wrench that tightens the terminal cap to secure the connector, the same wrench used to disconnect the MC4 terminals.

Chapter 9
How to Build Your Solar Panels Safely

If you are already planning to install your solar panels, keep these guidelines in mind so that you can follow them and be successful.

The Most Important Safety Rules

When installing solar panels, safety measures must be taken, from transferring a solar panel to starting it up.

While Moving It

There are no set rules for this section, but it is recommended that you make sure that all materials and products used in the installation are in their respective packaging, as they may suffer some external damage during the trip. If you use a liquid electrolyte battery, you must always drive upright.

In the Installations

In terms of installation, there are many standards for the installation and connection of electrical circuits, which can be found in the regulations of your respective country. It specifies how connections should be arranged and how they should be routed to separate and identify AC and DC circuits.

You must pay attention to the structure at this stage, especially the solar panels, so that they do not collide. It is best if you do the procedure orderly, i.e., install the structure first and then perform the module installation and electrical connection. All of the above factors allow the photovoltaic system to work, creating an order to secure the material and anticipate any electrical failure.

During Operation

The most important thing is to know which parts transmit electricity directly to understand how to manipulate the structure. Also, according to official standards, certain accessories should not be exposed to hostile environments, such as inverters, to protect your systems and anyone nearby.

You must ensure that the correct materials are used; for example, certain requirements must be met in wirings, such as resistance to a minimum of 600 V, resistance to high temperatures from the sun's rays, and compliance with necessary certifications. Yes, even the cables must be certified!

I recommend that any non-professionals in the area keep their distance from the device when making connections.

Have Safety Equipment

Just as installers must have certain skills and experience, they should always have safety equipment to protect your bodily integrity as there are many risks, such as burns, electrocution, falls, or fractures.

How to Work Safely with Electricity

Periodically inspecting the electrical installation at home helps to improve your home's electrical safety, whether on a day-to-day basis or when performing new installations or repairs. You have to make sure that the installation is in good condition and properly grounded, that the electrical installation is in perfect condition and that the grounding wire is correctly placed. You should wear suitable clothing, ideally tight-fitting clothing and non-slip rubber-soled shoes. It is important to do the following:

- Ensure that the area is completely dry.
- Do not wear metallic items on your body, such as rings, watches, etc., as we may inadvertently cause the current to deflect them.
- When repairing lights, plugs, outlets, or electrical installation components, make sure that the electrical current to the house is completely disconnected.
- If it is not clear whether the power is completely disconnected, don't risk it.
- Use the right tools for the job.
- Wear gloves if possible.
- If you are away from home for a few days, it is a good idea to completely disconnect power to the equipment, thus reducing the risk of short circuits and fires.

Safety Rules for Working on Roofs

Safety is very important when performing any type of work, even if it does not seem dangerous. Working on a roof is always risky and requires extreme safety measures, as any slip and fall can have serious consequences.

One of the most common ways to reach the roof is by using a ladder, but keep in mind that not all models work well. It is best to use a telescopic ladder. Once you reach the roof, wear a safety belt that fits your body and adheres perfectly to the safety guidelines.

Remember the following tips:

- Before climbing, find out about the roof's type, thickness, and pitch.

- Do not climb on the roof when weather conditions are not optimal (rain, hail, frost, or strong wind).
- Do not climb on the roof when recent rain could get you wet.
- Do not climb on the roof without proper footwear to avoid slipping or a possible fall.
- Do not walk on roofs without guaranteed minimum safety conditions.
- Do not go on the roof alone.

Preparing for Bad Weather

Rain has little effect on the performance of solar panels because they capture solar energy even when it rains. Solar panels are built to withstand adverse weather, and by their nature, they are outdoors, so you don't have to worry about that. The materials that make up solar panels are tough enough to withstand rain and even hail.

Many believe that hail can puncture solar panels and stop them from working. However, you should know that solar panels have a solid frame, which means exposure to these elements does not risk your investment.

As for rain, you can rest easy knowing that it won't damage your solar panel system, which will help keep it clean.

Some factors affect the way panels capture solar radiation. For example, shadows from trees or tall buildings next to your house.

Lifting and Handling Solar Panels

Starting with the packaging, it is important to follow the instructions for the maximum number of panels allowed. In more developed markets, there are specially designed containers for the transport of modules. It all depends on the brand and supplier, so it is advisable to follow the manufacturer's instructions. Since photovoltaic modules are considered fragile materials, all precautions must be considered.

In case of discharge, personal protective equipment should be used. Always carry the module from the aluminum frame between two people, making sure not to support it with another part of the body.

In addition to this, it is important to note that photovoltaic modules cannot be stepped on.

When anchoring to a structure, suitable fastening devices must be used and designed to perform this task.

It is important to handle the tools used to tighten the screws, avoid touching the module glass, and, most importantly, to not drop tools on them.

It is also necessary to know and apply the required torque for the screw; the manufacturer will indicate this value.

The connection of the modules should only use approved connectors, such as MC4, and the module wires should not be spliced in any case.

After making sure that the modules are no longer moved or manipulated, a connection must be established between them, carried out by trained personnel.

Weather conditions should also be considered at the module installation, as the installation is not recommended when storms or strong winds are nearby.

Conclusion

It may be expensive to install solar panels but keep in mind that manufacturers guarantee 25 to 30 years of usefulness, and they will surely last longer if you take care of them and give them proper maintenance.

As another option, you can also have solar generators, which will give you less utility, but they can still solve the essentials.

We need electricity. It is a fact of modern life, from telephones to computers, not to mention the stove, oven, refrigerator, and everything else that you have in the bunker.

Almost everything in our homes depends on electricity to function. A power outage can severely affect our lives as the tools and equipment we depend on become unusable.

A portable solar generator is a great tool to give you peace of mind while keeping your equipment and appliances running in an emergency. If you can have solar panels and a generator, all the better; you have more options and even have electricity if you need to move.

Made in the USA
Las Vegas, NV
13 October 2023